Sarah S Wolverton

Primroses

Sarah S Wolverton

Primroses

ISBN/EAN: 9783337420000

Printed in Europe, USA, Canada, Australia, Japan

Cover: Foto ©berggeist007 / pixelio.de

More available books at **www.hansebooks.com**

PRIMROSES

BY
SARAH S. WOLVERTON

BUFFALO
CHARLES WELLS MOULTON
1895

COPYRIGHT, 1895,
BY SARAH S. WOLVERTON.

PRINTED BY
CHARLES WELLS MOULTON
BUFFALO, N. Y.

TO MY DAUGHTER
AND TO THOSE
*FRIENDS WHOSE TOKENS OF AFFECTION,
AND WORDS OF ENCOURAGEMENT,
HAVE HELPED ME TO BIND MY*
"*PRIMROSES,*"
*I DEDICATE THIS SIMPLE HEART-OFFERING
IN GRATITUDE.*

CONTENTS.

	PAGE.
I'll Mark My Place With Flowers	9
Stir the Fire	10
The House of God	11
To My Opposite	12
My Souvenir	13
A Christmas Gift	14
It Makes No Difference Now	15
Inspiration	16
Longing	18
Pearls	20
A Mother's First Kiss	21
November Sweeter than June	23
Thou Hast Called Me	24
An Art Amateur	25
Love Stronger than Death	27
That Old Straw Hat	28
Dr. Richard Inglis	31
Trust	33
One of Many	35
Beautiful Islands	36
Sing On, O Bird	37
To Give Is to Live	38
Rondeau	40
Alone	41
Triolet	42
Disappointment	43
A Plea for Home	44
On the Birth of My First Grandson	46
The Ocean Monument	47
The Harp and the Winds	48
A Query	49
Gleams of Sunshine	50
Retrospect	51
A Reminiscence of Childhood	52
Thanksgiving	54
The Awakening	56
Sweet Rest	57

Rondeau	59
A Rift	60
Sincerity	61
Where Hast Thou Gleaned To-day?	62
Triolet	63
To Josie	64
A Tribute	66
In the Heart's Temple	68
Baby Boy	69
Owls Kill Humming-birds	71
Don't Look at the Cobwebs	72
Only A Word	74
More Light	76
Battle Hymn	77
Where Do the Pins Go?	79
Fulfillment	81
The Angel's Visit	82
Lend A Hand	85
Triolet	86
Anchored	87
Two Little Hands	89
Have Faith	91
Be True to Me	93
My Beautiful Star	94
The Tramp Transformed	96
A Lesson from A Clothes-line	98
Shadow Watching	99
Mindful of Little Things	102
Two Ships	104
At Rest	106
The Poet	108
Tissue Transcript	110
Sonnet	112
Grant	113
Witch Hazel	114
An Invocation	115
Bessie	116
Rondeau	117
Father, Guard My Boy	118
My Flower Boat	119
Miriam	121

PRIMROSES.

I gathered some at morning hour
  Their petals wet with dew:
And some were found at high noontide
  Dotting the woodlands through.

The sweetest when the sun rays touched
  With red the western sky:
I bind them now as twilight falls
  Primroses! who will buy?

MARK MY PLACE WITH FLOWERS.

TO H. A. C.

I SAT one morning with a book
    Abstracted, and alone,
And thoughts suggesting other scenes
    Across my mind were thrown.
I looked around to find a mark
    With which to keep my place,
But looked in vain, until I saw
    A well-filled flower vase.

I rose and plucked a petal fair,
    'Twas fit for Flora's bower,
And said, "there is no better way
    I'll mark it with a flower."
And then I thought, "just so through life
    While time my hand empowers,
Between the pages of each day
    I'll mark my place with flowers."

STIR THE FIRE.

WHEN the blues are all about you,
  Trials hard your spirit tire,
Take the tongs, and for amusement
  Just sit down and stir the fire.
First you make a careless motion,
  Then more mindful will you grow
Gleaning whole estates of sunshine
  From the firelight's fitful glow.

And you watch the flames go upward,
  Each intent to do its best,
Filling space that is assigned it
  Quite unmindful of the rest.
You will feel your heart grow lighter,
  Better thoughts will gather near,
And within that glowing splendor,
  Life more worthy will appear.

Often when the blues o'ercome us
  'Tis the weakening ague chill;
And 'tis useless then to argue
  "Rise above," "exert your will,"
Just sit down and stir the fire,
  Feel its warmth through every vein
Then sad thoughts and gloomy fancies
  Won't have half their power to pain.

THE HOUSE OF GOD.

"THE PURE IN HEART SHALL SEE GOD."

O, FATHER! make my mind Thy habitation;
    To me Thy presence oft reveal;
    With love my spirit seal.
Purge from my heart each sinful vain formation
    That it a temple purified may be,
    Fit place to worship Thee.

When through the flowery fields temptation's calling
    May I Thy silent whisperings hear,
    O, then to me draw near.
When gloomy shadows round my life are falling
    May I walk firmly Thine appointed way
    Thy smile my guiding ray.

Aye, though it lead where every light receding,
    Doth waken doubt and cause the cheek to pale
    As frightful fears assail,
Through sorrow's night, where hearts with grief are bleeding
    Beneath affliction's chastening rod,
    E'en there may I see God.

TO MY OPPOSITE.

O, DO not try! You can not walk,
    These untold ways with me;
Too oft the trials of the way
    Your partial eyes would see.
I'd skip the rocks to glean the moss,
    And make a pillow sweet;
But you the rough sharp, edge would note,
    And bathe your aching feet.

I'd watch the gleam in yonder sky
    The sun just breaking through;
The gathering clouds foreboding wrath
    Would deeply trouble you.
I'd quaff the wine from out the cup,
    And deem its color fair;
You'd drain the dregs to analyze,
    And find deception there.

Then do not try, but let me go,
    I walk not all alone;
A hand is ever holding mine,
    I hear a whispered tone,
A radiant light is leading me
    Unseen by mortal eyes,
But O, so bright, it makes this earth
    For me a Paradise.

MY SOUVENIR.

AH yes, 'tis priceless in its worth, a thing
    That like the wand magicians hold for me—
Turns night to day, and by its light I see
The way to walk whatever fate may bring,
And with that happy thought I sit and sing
    Such songs as thrill my soul with ecstasy
    While through my heart sweet memories wander free,
And in my lap their garnered treasures fling.

Did e'er the artist while his fingers wrought,
    Tracing the lines with patient loving care
    Suspect another greater yet than he
      Would leave inscriptions deftly graven there
    That through the years would precious tribute be
Of love to Love's own sacred altar brought?

A CHRISTMAS GIFT.

THOUGH hands and feet must quiet be
    The heart its impulse weaves,
And says: than naught 'twere better far
    To give a wreath of leaves.

Then take this book, and as thine eye
    Shall scan each lettered page:
Bid reason bide awhile away,
    And fancy bright engage.

With magic wand she'll change each word,
    Till all shall seem to be,
The sweetest token love could find
    And sent from me to thee.

IT MAKES NO DIFFERENCE NOW.

DID'ST ever low before a shrine,
    In sorrow's garments bow,
And ask that one sweet flower be given
    To gild the ways of now?

That gift witheld, did'st go away
    And press the flinty stone
And while the flowers bloomed everywhere
    Did'st weep that thou had'st none?

And sitting there hast had a friend
    Come softly stealing near,
To say with pleasant words and smile
    " Dry up the falling tear."

Then all too late that other hand
    Reached forth the flower, but thou
Could'st only breathe in anguished tone,
    " It makes no difference now."

INSPIRATION.

PUT away thy rule and line,
    Give free run to thought divine,
Take no plummet, only seek
Words of tender love to speak!
Some lone heart may catch their ringing
Blessing thee while thou art singing.

Fearest thou such utterance may
Cheer no pilgrim on his way?
    That thy singing may be vain
    To assuage an hour of pain?
Trust the voice so strong appealing
Trust thy soul its light revealing.

Hark! the wild bird trills its song
While the blue he skims along.
    See, the trees with changing leaves
    How their song oft interweaves.
Loving work are flowers doing
O'er the earth their beauty strewing.

Learn from these O poet learn
When within the fire shall burn

Thoughts then rise thou must express
Thou knowest not what heart to bless.
Learn from Him whose hand is guiding
What shall fail, what be abiding.

LONGING.

THE winter winds are chilly, dear,
    While all around is hushed and still
I wish that I a step could hear
    To stir my pulses' deeper thrill;
I would that I could summon thee
    With magic of the will.

The falling snow is drifting past,
    Its chilly flakes my sight oppress
Although I fear not winter's blast
    There comes a sense of loneliness;
I would that I could welcome thee
    My longing heart to bless.

Can'st thou not hear the echoes ring
    Through all the live-long day from far?
Nor see at eve first glimmering
    One lonely, loving little star?
For one alone it shineth there
    Across the sunset bar.

Listen! as eve's calm shadows fall
    In myriad shapes that softly glide,

A silvery voice will sweetly call,
   " 'Tis trysting-time at eventide!"
Look up! the stars are shining bright
   But one shames all beside.

PEARLS.

WHEN wild the winds are raging
　　The billows loudly roar,
Then lowly bend and seek thee
　　For pearls along the shore,
Think not the sand and rubbish
　　Are only barren ground,
For often in such places
　　The purest gold is found.

And when the waves run highest,
　　And wash the rocky strand
'Tis then the troubled ocean
　　Casts treasures on the sands.
'Tis then those fearless spirits
　　Who brave the stormy night,
And walk within the tempest
　　Find gold and jewels bright.

And so when God is holding
　　His chastening rod above,
'Tis they who bend in meekness,
　　Who win His precious love.
'Tis they who walk with patience
　　Along life's troubled strand,
Who meet the holy angels,
　　And clasp the Master's hand.

A MOTHER'S FIRST KISS.

SEEKER of beauty, arouse thee! away!
    Bring me rich gifts for my love I pray,
High up as the azure, deep as the sea,
Search for their treasures, bring them to me.

I scaled the vast mountain, traversed the mine
To bring thee rare gifts in a crown to shine.
Up through the cloudrift, cleaving the blue,
Diamonds I cut from the crystallized dew,

From silvery threads of the moon's soft hair
Wove I a robe for a bride to wear;
I fathomed the ocean, found 'neath its wave
These corals adrift from a mermaid's cave.

I found on the waves of the wind afloat,
The merriest song from a wild-bird's throat,
And flowers bathed in a sunbeam's kiss
I gathered to grace thy bower of bliss.

The treasure most valued I could not obtain;
I bartered, I bargained, and pleaded in vain.
On affection's warm breath the treasure had shone,
But e'er I could grasp it the tribute was gone.

But memory holds, forever will hold,
The glimpse of that jewel more precious than gold,
Wait you for my answer? It only is this
On the lips of a babe, a mother's first kiss.

NOVEMBER SWEETER THAN JUNE.

LIFE is not always sweetest when the dew
    Beads on each leaf and tree and fragrant flower,
  Sometimes there breaks a glad autumnal hour
That sweeter is than summer ever knew.
In life's young day we fancy all things true,
    Imprisoned in her flowery, fettered bower
  Imagination then wields wondrous power,
And through her lens we every object view.

But in November days, upon the field,
    Where once we plucked the roses fair of June,
  The richest fruits ungathered oft remain
And with their splendor all the ground is strewn.
'Tis so with life, the most abundant yield
    We find unlooked for on its western plain.

THOU HAST CALLED ME.

THOU hast called me thine angel
　　When sorrow was nigh,
No cloud rift to lighten,
　　No star in the sky.
Thy feet were aweary
　　Thy heart beating slow
But swift through the tangles,
　　Thy footsteps must go.

I came as a flash comes,
　　I spread the white wing
Through the mist and the darkness,
　　Glad tidings to bring.
To fold the loved wanderer,
　　Close, close to my breast,
To soothe the worn spirit
　　There sweetly to rest.

And still as we journey,
　　The blue hills below,
To cheer and to comfort,
　　The white wings will go:
Their sheen may be hidden
　　No sign shalt thou see.
But surely beside thee,
　　Thine angel will be.

AN ART AMATEUR.

MRS. BOBBY TALKS ABOUT "PAINTIN'."

BEHOLD how great a gift it is
    A gem like this to paint!
Some people have the genius in
    And some there be that ain't.
I never guessed I had it in
    Till Betty brought it 'eout
-And now there ain't a airthly thing
    I care so much abeout.

At first I used to nuss a bit
    To go of errants too:
But somehow that wa'rnt quite enough
    For willin' hands to do.
At first I took to paintin' fruit:
    Pears, peaches, and the like,
But neow to grander spheres I soar
    For higher Art I strike!

To landscapes now my soul aspires,
    I'm bound to put her through,
To paint trees, rocks, and rivulets
    Some skies and mountains too.

I'll climb atop the hill of fame
I'll fling my brush in air:
I'll plant my easel in the clouds
And paint a picture there!

LOVE STRONGER THAN DEATH.

I THINK if I were dying and you came
    And took my trembling hand
And waited by me on that dreary brink,
    They call the Border Land :

And said in your kind gentle voice, " cross not
    That rapid river, dear,
'Twould darker grow if you were gone, we need
    Your presence daily here."

I think I'd know the hand, and, that new strength
    Through every vein would thrill :
While from the soundings of my heart would wake
    To life the sleeping will.

I think my soul would stay its flight nor care,
    Although I'd wandered far,
Though gleamed the City just before, and wide
    The golden gates ajar.

Nor question aught of good or ill, I'd know
    Joy in ourselves is found :
That if our presence lendeth light, then will
    Heaven lie all around.

THAT OLD STRAW HAT.

SEE! here is the house, 'tis old and gray
That is the cottage over the way.

He lived in the cottage, then 'twas white,
I in the house, "old Pease" color quite.

"Old Pease" was a Quaker, stiff and staid,
And all alike were his houses made,

And just the same were they painted too,
His coat and hat they matched in hue.

The roofs were red, and like a star
Each beamed upon you, near or far.

Well many a time e'er the morning sun
To waken the neighbors had begun,

I'd listen a well-known whistle to hear
Drawing my window softly near.

Then down the stairs and out of the door
I would skip the green grass gaily o'er,

To catch the gleam of a bright brown eye
As he tossed his straw hat up to the sky.

Then, where the meadow pathways wind,
The daisies we in chains would bind.

Buttercups in their hearts of gold
And, "all for you" his brown eyes told,

And berries ripe—there are none to-day
As sweet as those on that highway.

And roses fair, though they'ro fragrant all,
Grow fairest yet by that old stone wall.

And then at eve as the sun went down,
His path to the gate my whistler found ;

Again to the meadow would we roam
To drive the cows in safety home ;

Then sit on the stile and stories tell
Till over the hill long shadows fell.

Nor little dreamed—either he or I
How soon we would say our last good-bye.

I went to the city. On sped the years,
Bringing their weight of joys and fears.

A woman's wish and a woman's pride,
Children now were my steps beside ;

I saw him once again—Ah me,
I naught of his boyish look could see;

A stalwart man with a bearded face,
Of his youthful self no lingering trace.

A stranger now, by the gateway he
No longer whistles and waits for me.

But memory hangs in her silent hall
Beautiful things at the spirit's call;

Landscapes, whose colors deftly laid
By unseen artists, can never fade.

And one, O fondly I gaze thereat,
The face of a boy in an old straw hat.

DR. RICHARD INGLIS.

DIED 1875, AT DETROIT.

" For he came not to be ministered unto,
but to minister."

AND so through the paths his Master trod,
    Where the sick and suffering lay,
Himself forgot in his glorious work,
    He journeyed on his way.
His calm, clear eye in the distance saw
    Where the noblest heights upreared
And his daily life of doing good
    Like a hallowed round appeared.

O, not for gold was his service wrought,
    The poor man's grasp was as dear :
The feeblest wail from an infant's lip
    Fell sad on his listening ear.
The moan of pain from an humble cot
    To him was summons as great
As that which called to the princely home,
    By luxury's couch to wait.

Aye, just as warm was the clasp of his hand,
    His smile would as sweetly break,

His voice was as soft and tender toned,
   While comforting words he spake ;
The stranger he met beside the well,
   With a life-load hard to bear,
Would look in his face and find a friend
   And comforter written there.

While the young untried, just entering paths
   Where way marks are often dim,
Found ever a guide to counsel, direct,
   Through every trouble in him.
He has gone from our sight, his life-work is done,
   But others are going his way.
The record he leaves, O, let them take up
   And practice his precepts each day.

TRUST.

FALLING, yet falling o'er my life
    Are shadows dark—but still
I whisper, "there is One above
Who watcheth o'er his child with love,
    All things obey His will.

He will listen if I call him,
    And send an angel down
To guide me as the night comes on—
To lead me when the lights are gone,
    "Through ways I have not known."

He will not give a heavier load
    Than I have strength to bear:
And if I climb the mountain steep
His eye a kindly watch will keep
    Where thorns and rocks appear.

In Him I trust, on Him I lean
    Through all life's devious ways,
The past, the present, future—all
Are His: and when the shadows fall,
    His are the cheering rays.

His angels trim a golden lamp
   Above the cloudy sky,
There brighter far than sunny beams
Upon life's darkest hour it gleams
   Before faith's trusting eye.

ONE OF MANY.

AH no, to thee I would not be but one
    Of host innumerable ! I would be
The one best loved, and I would have thee see
Each day my face as sees the morn the sun
Were pathways rough and drear for thee begun,
    As to the light, thine eyes should turn to me,
    That I might cheer or comfort give to thee
Until the care and weariness were done.

My heart should be thy resting-place, and I
    Would find my joy in knowing that howe'er
This life might measure out of good or ill
    Through every mete none other came so near :
No other hand thy cup so full could fill
    And change to sweet, even Marah's waters drear.

BEAUTIFUL ISLANDS.

THERE are beautiful isles in life'e ocean
    With the choicest of flowers o'erlaid ;
We will find them where deepest affliction
    A path for the traveler has made.
By the ploughshare of sorrow the furrows
    Are turned on the smooth waiting sod,
While the seeds of the blossoms unfolding,
    Are sown by the wisdom of God.

Dear feet seek those beautiful islands,
    Dear hands of their treasures oft glean,
While the hearts that are tender bend o'er them,
    And love adds delight to the scene ;
From the ships that glide by on ocean
    The sailors keep watch for their light,
They suffer no danger from shipwreck,
    With those beautiful islands in sight.

The aged ones tottering onward
    Clasp closely the hands that are there—
The mourner is constantly seeking
    Assistance his burden to bear.
Though the sunshine may sometimes be shrouded,
    Yet softly the shadowings fall,
For over those beautiful islands
    Reigns God, the Commander of all.

SING ON, O BIRD.

SING on, O bird, thy mate to cheer,
    Thy sweet love-notes she waits to hear:
However high may be her flight
With wings a-droop she will alight,
Straining a loving, listening ear
To catch the song to her most dear.

Though heavy hang the clouds and drear,
    Or breaks the golden morning light,
        Sing on, O bird !

For one sing on ! Sing loud and clear
Through all the day till stars appear.
    Let no cold winds thy voice affright
    From tallest tree-top's dizzy height,
'Mid foliage green or leaflets sere,
        Sing on, O bird !

TO GIVE IS TO LIVE.

RESTING one time on my journey
　　Close to a circling wall,
Saw I in fancy at twilight
　　A pine, shapely and tall.

The foliage of autumn enclothed it,
　　Covering its branches wide;
Only that one in the meadow,
　　With waving, low grasses beside.

And ever it seemed to be sighing,
　　Making inaudible moan;
Something within it kept saying:
　　"Forever, forever alone!"

Down at its base in seclusion
　　Nestled a lone little vine,—
Looked up at the glory above it,—
　　Cared for the heart of the pine,—

Looked down at its tendrils trailing,
　　Broken by showers,—
Thought of the fibers frail, forming,
　　Folding the flowers.—

To Give is to Live.

Smiled to itself in its pleasure,—
   Crept close to the pine;
One after another, its tendrils
   Began to entwine.

Up through its outermost branches
   The pine to its crest
Stood freighted with blossoming beauty,
   Down into its breast.

No more, though falling the twilight,
   No more, as a call for relief,
Heard I that innermost mourning,
   That echoing accent of grief.

Strength for support in its clinging
   Perfected the vine;
The tendrils so tenderly twining
   Gave joy to the pine.

RONDEAU.

LOVE fell asleep long years ago:
   I let him sleep: I did not know
  How I should care his smile to see
  Beam bright one day on me.
How could I tell it would be so?
Beside the road where briars grow
And bitter, biting, north-winds blow
  And darkness gathers dismally
    Love fell asleep.

But ah! one day—oh Time be slow!
Haste not Life's triumphs—soon they go,
  Love woke! He clapped his hands in glee
  Cried out as one from prison free:
"I did not die!"—his cheeks aglow,
    "Love fell asleep."

ALONE.

"ALONE!" oh no, I'm not alone
   A gentle spirit walks with me
And though the shadows gather dark
   Its loving smile I still can see.

It wreathes my brow with memories fond
   When racking pain waits me beside
And through all griefs to heights unknown
   Leads me that tender spirit guide.

And when my footsteps falter oft,
   As faint, I scan the darkness wild
It whispers to me, " fear not thou
   God watches ever o'er his child."

And oh, the visions it will bring
   For me, to show the perfect bliss
Awaiting in the world above
   When I released shall rise from this.

I see the pleasant pastures there,
   I hear the flowing crystal tide
Where safe from all I've suffered here
   I rest among the purified.

TRIOLET.

IF I could tune my lyre, dear,
   To notes the gracious angels sing
No saddened strains should greet your ear
If I could tune my lyre, dear,
Each note should wake so sweet and clear,
   The music peace to you would bring,
If I could tune my lyre, dear,
   To notes the gracious angels sing.

DISAPPOINTMENT.

I'VE watched the morning o'er the hills,
    Bright evening turned to gray;
And come and go the happy time
    Of Merry Christmas day.

And yet the friend I fondly hoped
    Upon that day to greet,
Gave not the tender hand to clasp,
    The Christmas welcome sweet.

Ah me! Life seems a sea of gloom,
    The crossing ways are dark
And out beyond the beacon lines,
    Alone my little bark.

Alone? above are steady gleams
    To light the gloomy sea,
And there are spirit-friends I know
    To guide my course for me.

They loved me once—nor can it be
    They'll let me come to harm,
But lead where disappointment ne'er
    Can give my soul alarm.

A PLEA FOR HOME.

"We may build more splendid habitations
But we can not buy with gold the old associations.'

MY home, oh do not take it,
   'Tis all the world to me!
Tis full of hallowed memories,
   My loving eyes can see

Each cherished chair and table,
   Each book and tinted view,
The brackets and the vases
   Bear written records true.

Each has a clear inscription,
   The name I treasured so,
Of some dear one who loved me
   In days of long ago.

The carpets trod by strangers
   Are not mere woven thread;
They bear the print of footsteps,
   Long silent with the dead.

The walls resound with voices
   That still to me are dear,
When through the twilight silence
   Their loving tones I hear.

Then memory brings the faces
  On which I love to gaze,
To learn the worth of living
  In Life's deserted ways.

Oh ask me not to leave it,
  'Twould break my clinging heart
To tear its loving tendrils
  From that dear home apart.

ON THE BIRTH OF MY FIRST GRANDSON.

FAR, far away over the mountain
   In a snug little valley below,
Where a creek through the canyon is winding
   And lightly comes falling the snow,
Is a ranch. There are father and mother,
   A daughter as fair as a rose,
That a brightness like sunshine all over
   The home of the wanderer throws.

But with all, there was ever a longing,
   A yearning and coveting there
For a son to come into the household
   And a place in its heritage share.
Till one morning just after Thanksgiving
   The signal was given to rise,
For the angel of Life was approaching
   And bore in his bosom a prize.

Soon a dear little mortal was cuddled
   Close down in its foldings of white,
And a baby boy opened his eyelids
   To gaze on this wonderful light.
Then a messenger swift as the lightning
   Sped over mountain and sea—
Abroad on the wings of the morning,
   And brought the glad tidings to me.

THE OCEAN MONUMENT.

O, BRIGHT the ocean monument,
  The star-reflected light,
Above the lonely sleepers through
  The long, long, stretch of night.
The wind in wild sweet melody
  An anthem loves to play,
Reminding of the sleepers there
  Through all the lonely day.

O, Ocean's dead shall never be
  Forgotten or unknown,
Because no names are registered,
  Upon the sculptured stone.
Because no weeping willow waves,
  Its branches, green and fair,
The starlight and the wild-wind harp,
  Keep constant vigil there.

THE HARP AND THE WINDS.

'TWAS a beautiful harp, but no right hand
    Had over its life-chords swept;
And through the years closing it in,
    In the hall it quietly slept.

But the south wind came, his touch was warm;
    His wand o'er the wires he flung,
When the strains burst forth, as an anthem sweet
    To the voice of the Master sung.

Then it trembled, thrilled through every chord
    With new life waking within,
As you've seen a bush when a flock of birds,
    Just alight 'mong its boughs has been.

But the north wind rose in his region of snow,
    Bore down o'er the hills in his pride;
When the soft south wind sank down, and low
    In the cold and the snow sheets, died.

You must know how the great dark settles in?
    Sometimes not a sign nor a token,
How the stars go out, how the moon is not?
    It is so when some ties are broken.

A QUERY.

DID it possess a deathless soul,
    That little one of mine?
Who just passed through the door of life,
    Was gone at day's decline.
Its close-shut eyes had opened not
    Its hands so pale and cold,
Gave back no pressure when I warmed
    Them in my bosom's fold.

Its little feet like pearls were hid
    Beneath its robings white.
Ne'er were disclosed those baby steps
    To glad its mother's sight.
So small, so frail, and yet my heart,
    Felt sorrow's rushing tide
When kind hands brought the babe to me,
    And laid it by my side.

And great the love bestowed on me
    When in a lonely hour:
I heard the whispered promise made
    To give a precious flower;
And so by that I know it lives,
    And waits for me above
And reaches down its baby-hands,
    To plead for mother's love.

GLEAMS OF SUNSHINE.

YES, there are glimpses of sunshine,
   No matter how heavy the cloud,
Stars in the firmament shining
   The brighter that shadows enshroud.
There are words that are silently spoken,
   Whose echo rings loudly and clear
When high over tones of contention,
   Their comforting music we hear.

There are smiles like the blossoms of summer
   Erasing the up-springing frown,
The sweetest when trials are frequent,
   The brightest when fortunes are down.
Caresses we always remember,
   That soothed us so softly to rest,
The peace of a beautiful nature,
   That gave us its truest and best.

The clasp of a hand strong and tender,
   That always imparted delight:
We knew that though hidden the pathway,
   It ever would lead us aright.
Good gold that we never will barter,
   Great treasure we never will lend,
The jewels and gold of that sunshine,
   Where love and adversity blend.

RETROSPECT.

I'M looking fondly backward
   And living o'er again
Some cherished scenes that memory,
   Has pictured on my brain.
I'm thinking of a morning
   Not many months ago,
When some few words were spoken
   In tender tones and low.

When danger had been threatening,
   And I was crushed with dread,
You placed your arm around me,
   These very words you said:
"Come, tell me all about it!"
   I hardly then could speak,
The fright from which I trembled,
   Had made each accent weak.

My soul grew brave and stronger,
   For something in your tone
Conveyed the sweet impression,
   I suffered not alone.
The years may come and bring me
   Words spoken kind and true,
But none will seem more precious
   Than those I heard from you.

A REMINISCENCE OF CHILDHOOD.

IN looking o'er the landscape
   As scenes pass to and fro,
Comes floating in the window
   A leaf from long ago.
I see the dear old homestead
   The childish faces there,
And grandma sitting smiling
   In her old easy-chair.

I see the flower-garden
   Where bloomed the lilacs sweet,
Along the walks impressions
   Of many children's feet:
I see the mimic garden
   Beside the rocky wall,
One lady-in-the-bower
   And one blue larkspur tall.

While all around the border,
   There stands a fence of pine
By childish hands constructed,
   And given me for mine,
A cistern by the platform
   Where burdock leaves abound

And where my precious dollies,
  The ragged ones—were drowned.

Ah me! how fell the tear-drops
  Down childhood's tender cheek,
My grief was then as bitter
  As woman's tongue can speak.
When home from school returning
  That cistern met my eye,
My dollies in it hanging,
  Their feet turned toward the sky.

THANKSGIVING.

THE quiet guests assemble,
  The guests from far away
Come with their silent greeting
  On this Thanksgiving Day.
Down through the realms of memory
  They come on every side,
Within my heart's great recess,
  A little while abide.

One from her Southern homestead,
  In accents soft and low;
Says, "Dost remember dearie,
  Thangiving long ago?
While friends were gathered round me,
  That honored day to keep,
Without a word of warning,
  I laid me down to sleep!"

Another loved more dearly,
  Clasps close my willing hand
And—these the words that whisper
  From out the Silent Land.
"Thanksgiving dost remember
  It seems so long ago

Thanksgiving.

The day was like to this one
   As softly fell the snow?

" You could not tell the meaning
   Of gloom that round you stole
Or why the shadows sombre
   Enshrouded all your soul.
But few the weeks had numbered,
   When you 'mid sorrow's swell,
Remembered that great darkness
   And knew its import well."

And o'er the snow-clad mountain
   Fond memory brings her own,
While loving arms and tender,
   Are fondly round them thrown;
And so though uninvited,
   The guests who grace our board,
The loved ones who are vanished,
   Are to our hearts restored.

THE AWAKENING.

HER harp stood silent in the hall. None, none
    To see, to test the fitness of its string,
    To guess the power it held within. Or ring
From out its depths the melodies sweet: not one.
But ere the spring had caught the summer sun,
    One came. He saw, he bade her bravely fling
    Her hand across the chords and to them bring
The timid music o'er the bars to run.

'Tis autumn now: those wakened chords no more
Are mute: beneath the touch of love's soft hand
    Enrapturing all with song, her senses thrill,
I wonder if within a distant land,
    Those notes transplanted can one bosom fill,
    Can charm where she the harpist must adore!

SWEET REST.

AT last has come to thee that precious boon
    Long sought, sweet " Rest."
I wonder as I read those lines, what cross
    On thee was pressed,

That thou should'st long to have life's journey o'er
    And be at rest.
I know some ways are rough: that paths divide:
    Where we think best,

Our father bids us not go on. Oft too,
    Where flowers grow
We ask with pleading tears awhile to walk,
    He says, not so.

I know we strive our burdens hard to bear,
    Strive to be brave!
To walk with loyal hearts, the weary path
    On to the grave.

I know that human strength is weak—that none
    Of human kind
Can take the burden we are called to bear
    By Him all-kind.

We'd weary of the road, but He who notes
    The sparrow's fall
Will lead through tangled wood or dark morass
    Past dangers all.

I too have cried for "Rest," and still I cry
    For I am weak;
And may not gain however hard I try
    The goal I seek.

Yet glean I in the roughened paths I tread
    Some flowers fair;
And try to catch the birdlings song, that greets
    Me everywhere.

And so I win my joys through life, and try
    To be content,
Believing this: or cloud or sun, each was,
    In wisdom sent.

And when the day is o'er, and sinks the sun
    Low in the west,
I hope to hear above the dark, "Come thou,
    And sweetly Rest."

RONDEAU.

WALKING the fields one summer's day
    We gathered flowers the seeds to strew,
We pulled the blooms but kept a few
Strolling along our pleasant way.
The sun flung down his golden ray
    And we were wrapped in light, we two,
        Walking the fields.

We walk again where sunbeams play
    The same old paths with pleasure through:
    New flowers are where the old ones grew,
We pluck them fresh in glad to-day
        Walking the fields.

A RIFT.

THE gloom had gathered close around
    Each ray of light obscured,
Submission only meant to be,
    I patiently endured.

When lo! the gloom began to part
    The cloud to break into;
A silver shinning prophesied
    A sunbeam skipping through.

A cheerful voice invoked the same,
    As tripping feet came near—
While smiles within a darkened room
    Made brighter all appear.

God bless the careful comforters,
    The helpers o'er life's way:
And as they measure may life mete
    Where'er their feet may stray.

The others? Ah! they have their own,
    Our pity they should share;
They miss one half life's blessedness
    The other can not spare.

SINCERITY.

JUST the words as nature gave them,
    Full or scanty, be my praise :
As the woodland birds are singing,
    I must trill my simple lays ;
From within, and not from seeking,
    Out among the trodden lanes,
Often comes the choicest offerings,
    Purest thoughts the mind obtains.

Each and all, we have our mission ;
    Some to strike the lowly lyre,
Some to stand on Mount Parnassus
    To invoke the fitful fire.
Crave I not that height oft lonely,
    Where the chilling blasts abound ;
In the warmer sheltered valleys
    Would I rather far be found.

Where the heart with heart is speaking,
    Where the friendly hand is warm,
Where we comfort one another,
    Should the shadow bode a storm ;
There in nature's sweet communion,
    Fond of heart, and pure of soul,
I would find my noblest mission,
    While the years their records roll.

WHERE HAST THOU GLEANED TO-DAY?

THE sun sinks low in the purple west,
    The birds to their nests hrve flown,
The flowers fortell the coming of night,
    Now, where hast thou gleaned, mine own?

Were those fields all of ripened grain,
    Didst bind up thy sheaves with care,
Were sorrow and pain, subdued and stilled,
    Did kindness reign ever there?

Didst buoy with hope the tempest-tossed one,
    And point to the beacon-light,
When rough the billows, and some poor soul
    Was caught in the chill of night?

Didst lend thine ear to the shivering cry,
    Didst give to the famishing, food,
And feel at heart ; for earth's poor ones
    Thou hadst done what thou could?

Thy grain is good, then, and God's own hand
    Shall garner it all for thee,
For separate on the great threshing-floor,
    The grain and the chaff will be.

TRIOLET.

CATCH ever the sunbeams, my dear,
    The sunbeams, though ever so small.
It matters not where they appear,
Catch ever the sunbeams, my dear,
Quite often a shower hangs near;
    Then, ere the rough rain-drops shall fall,
Catch ever the sunbeams, my dear!

TO JOSIE.

ON HER MARRIAGE DAY.

GOOD-BYE, my precious darling.
   From out the port of home,
Upon an untried ocean,
   Thy little bark must roam.

The stranger's hand must guide thee
   To fairer, sunnier clime :
His heart thy helm and compass
   Through all the future time.

Oh, may his hand be tender,
   Whate'er life's promise bring.
Oh ! may his love prove faithful,
   But brightness round thee fling.

Through joyful days or sad ones,
   The sweetest sounds to thee
His words of fond affection,
   In soulfelt sympathy.

But dearest, through the moments,
   That come all flower-crowned,
Sometimes let thoughts of mother,
   On folded leaves be found.

And when for thee life's shadows
  O'erwhelm thy spirit, dear,
Somewhere, they're sure to meet thee,
  In sombre garb, and drear;

Then, darling, call for mother,
  She'll bid each duty bide,
Take all thy heavy burdens,
  Or share them at thy side.

A TRIBUTE.

*" Earth would never touch her worst
Were one in fifty such as he."*

'TWAS night ! I wondered would I breast
    The wintry wave ;
Suppose a storm arose, what hand
    Was there to save.

Spake low a voice within—" Always,"
    Is One, He hears
His children's cry ; "the widow's God"
    Calm thou my fears.

Ah, yes ; I know. But I am weak ;
    I crave a hand
To outward reach : a human stay
    Of earthly strand.

Do angels listen ? Bear away
    Our every thought ?
And hovering near are angels too,
    With blessings fraught ?

I can not tell ; but out beoynd
    The sunset far,
With fervent, loving clasp, and clear,
    Across the bar,

There reached an earthly hand.   The years
    Fold over then,
But still a hand as kind, as firm,
    As true as when

The dark veil fell around that hand;
    A brother's clasp
Has bade me know how strong and pure
    Is friendship's grasp.

I think God works on human lives
    Through human means,
And what men call but kindly acts, '
    God's method screens.

IN THE HEART'S TEMPLE.

I BOW me not at fashion's smile,
   No homage give to gold,
But at the bar of intellect
   My casket open hold.
I bend the head, to wisdom heed,
   Wait breathlessly and still,
Lest by mischance a gem escape,
   Ere I my measure fill.

And where some gentle, earnest soul
   Puts self beneath its feet
To help to bear another's cross,
   And yield an influence sweet.
I love to bring my little mite,
   The robe of light to see,
And let perchance its outward reach
   Rest evenwhile on me.

Yet do not think I beauty scorn;
   Ah! no, she holds her own,
And everywhere my life lines lie,
   I like her emblems strown.
She bathes my heart in sunny rays,
   And keeps its temple fair,
And welcomes every genial thought
   That seeks an entrance there.

BABY BOY.

MRS. A. E. B.

ANGELS bright were floating softly,
   Through the fleecy clouds above,
And they often bent to listen
   For some whispered prayer of love.
Upward rising on a zephyr,
   Breathed a gentle mother's prayer :
" One more lamb within my bosom,
   One more birdling for my care.
Years ago to me was granted
   One sweet lamb, my arms to hold.
But our Shepherd saw and loved it,
   So he called it to his fold.
Though two others play about me,
   Bringing gladness like the sun,
Still dear Father, bend and listen ;
   One more like the absent one.

Those bright angels upward bearing
   That fond prayer the azure trod,
Till they neared the height celestial
   And they reached the home of God.

Down before the footstool bending
  With the host awaiting there
In its words, all pure and holy,
  Low they laid that mother's prayer.

Summer flowers blossomed sweetly,
  And the song-birds warbled free ;
Then thy prayer of love was answered
  One more lamb was sent to thee.
But my friend, remember ever
  It is but a treasure lent ;
And should God, the Shepherd, claim it,
  Thou must give a calm consent.
But I trust thy cup is measured
  Now to hold its fill of joy,
And that bright will be the future,
  Both for thee, and Baby Boy.

OWLS KILL HUMMING-BIRDS.

CRITICS when the birds are winging
  Through the air—their carols singing,
Peace and joy, to mortals bringing,
    Do not criticize.

While you may be very clever
Seeking for some blemish ever,
You o'erlook the sweet endeavor
    And the beauty dies.

O'er the world some hearts are aching,
Some in silence slowly breaking
Just because of people making
    Use of words unkind.

While we con the simple reading
Let us scan the magic leading
For the timid songsters pleading
    "Owls kill humming-birds."

DON'T LOOK AT THE COBWEBS.

LYING one day on my wearisome couch,
    Quite free for the time from pain:
My eyes roved around in search of food
    To nourish my hungry brain.

Cobwebs I saw from the ceiling hang
    In their tissues soft and thin,
Yet strong enough to easily hold
    The dust they gathered in.

It was not a sight to please the eye
    When the hands were helpless laid,
Nor yet a sight for the throbbing brain,
    That could give the hands no aid.

So I turned my eyes to another place,
    Where rested a picture fair,
Where the artist's touch gave flowers and trees
    And the bloom of summer there.

And finding food for the waiting brain
    In the bright beams floating by,
I quite forgot there were cobwebs dark,
    Still hanging my pictures nigh.

Don't Look at the Cobwebs.

And then I thought, how this life hangs full
   Of cobwebs close to the wall:
There are sights and sounds disturbing us
   We would gladly escape them all.

But as our feet must oft tread the path
   That the most discomfort brings,
It is best to turn from cobwebs dark,
   And look for the brighter things.

ONLY A WORD.

'TWAS only a word—but O so fond
It fell on a waiting ear,
That ever I knew
Though time may go
That sound it will gladly hear.

It wafted across the waters past,
Turned back the out-going tide,
It touched a bell
In memory's cell
Flung open her portals wide.

And the years long past seemed few and near
By memory's enchanted light,
As forms of old
Now still and cold,
Come back to my longing sight.

I trod the paths where the childish feet
Life's rounds would merrily go;
The merest dream,
A joy would seem
As gems in their luster glow.

And words I heard of the sweetest praise,
    That comes to a childish heart,
        Till tears that rise
        To a woman's eyes
    From my drooping lids would start.

That little word I have laid away,
    To me 'tis a beautiful spell
        For when alone,
        The star-light gone
    It has a story to tell.

MORE LIGHT.

MORE light! O Father, lend me more,
   'Tis not so dark I can not see,
But oft the shadows gather o'er
When far-off seems the better shore
   While gloomy doubt will nearer be.

I falter on my toilsome way,
   And scarcely dare my feet to move,
While veiled appears each kindly ray
That used my trembling steps to stay,
   When walking toward the heights above.

Faith's lamp oft fails me, burning low
   As if 'twould quench its flickering light
And shadows then will longer grow
And slower too my steps will go,
   I fear the shades of gathering night.

O, Father! let an angel come
   My little lamp of faith to trim;
Lest, 'mid the shadows as I roam
I stray so far from Jesus' home
   I'll ne'er behold the face of Him.

BATTLE HYMN.

"And the chariots of God are the thousands that move as angels to bless mankind."

ONWARD sisters! do not falter,
    Though the waves run mountain high:
You shall cross the Red Sea over
    And your sandaled feet be dry.

Fear not discord nor contention,
    They shalll flee as you pursue,
And the sneers that follow after,
    Sink into oblivion too.

Proudly, firmly, plant your standard
    Bid the world endorse each line,
Woman's wit, and woman's wisdom
    With the laurel wreath entwine.

Flash of sword, nor roar of cannon,
    Heraldeth our army's tread,
Leave we neither dead nor dying
    To proclaim a victory red.

Peaceful homes, and pleasant faces,
    Greet us wheresoe'r we go,
Brighter days through our endeavor
    Over darkened lives shall glow.

Onward sisters! never falter,
   Though the waves run mountain high,
You shall cross this Red Sea over
   And your jeweled feet be dry.

*   *   *   *   *   *

I could tell of forms now vanished,
   Lonely graves oh, far away,
Holding buds of precious promise
   Early fallen by the way.

I could tell of hours of weeping,
   Nightly watchings all alone,
I could point where one sat waiting
   Till the rays of daylight shone.

And the picture never fadeth,
   Though the years on years go by;
While I live and memory lingers,
   Leaves it not my childish eye.

So goes out my heart amongst you
   And I pray that God's right hand,
Lead you on to heights victorious,
   Drive intemperance from the land.

WHERE DO THE PINS GO?

I PLACE on my cushion a surplus of pins
   And say to myself, "these will last."
But not one can I find, according with mind
   Ere twenty-four hours are past.

The cushion is nice; embroidered in white,
   All daintily ruffled around,
No better a bin, for a good little pin
   In city or town could be found.

And yet in a hurry, if needing a pin,
   Its up-stairs and down I must run,
With jacket askew, my collaret too,
   In a chase of the kind there's no fun.

Now where do they go? There's no place above,
   In the measureless cushion of blue
Where pins, crooked or straight, may find their estate
   And carve out a destiny new.

Or do they go down and under the sea,
   Where mermaids their toilets prepare,
Come happily in to fasten a fin
   And find opportunities there?

Well this is a point that we're looking up yet,
  And the sharpest that carries a head,
May puzzle his brain again and again,
  A light on this problem to shed.

FULFILLMENT.

STAR-EYED and tender, a tiny flower,
    The blue veined forget-me-not
I saw, as I passed a floral group,
    In a lovely garden spot.

Softly folded the buds were waiting
    To open another day,
For me they held tender promise,
    I heard them all sweetly say.

The morning was gently breaking,
    My buds were opening there,
The folded promise had kept its faith
    And its fragrance filled the air.

"Ah, heedest thou dear, the lesson
    Written on a flower's heart?
Be faithful to thy promise ever,
    And perfect the smallest part."

THE ANGEL'S VISIT.

WHEN the summer sun was sinking,
    The flowers blooming free,
The wild bird's note resounding
    From every leaf and tree,
An angel bearing promise
    Down through the azure came,
And pausing at the portal
    He softly called my name.

I rose to do his bidding,
    He only whispered low
That when had passed the autumn,
    And came the winter's snow
He'd bring a precious flower,
    And place it on my breast,
And if came care or sorrow
    'Twould lull them into rest.

O, how I watched his coming,
    And waited at the door—
I knew he'd not deceive me—
    Three times he'd been before.
At length when evening's shadow
    O'er hill and valley fell,
We heard the white wings rustle,
    And knew the import well.

The Angel's Visit.

Within his hand he held it—
  That flower so pure and bright,
That all around seemed morning
  Though gathering near was night.
I reached my hand to take it,
  Glad joy o'erspread my heart,
When—was it true, or fancy?
  A something made me start.

Then back the blood went coursing,
  While stifling came the breath,
Drew near a dreaded presence—
  We felt that it was death.
He snatched my precious blossom
  From out the angel's hand,
And in his cold arms bore it,
  Unto the hidden land.

A light within is burning
  That gives a cheerful ray,
And through its soft, pure radiance,
  I trace my darling's way.
It passed where sin and sorrow,
  Sharp pains, and trials grow
And entered where no suffering
  The inmates ever know.

That City is Celestial—
  A world of untold bliss,
Where scenes of joy awaiting
  Atone for grief in this.

And where the fairest flowers
For life's brief hour made,
Are cared for by God's angels,
And never, never fade.

LEND A HAND.

*" Break not the bruised reed,
Quench not the smoking flax."*

WHEN the feeble frame is bending
   'Neath a load too great to bear,
While the footsteps linger weary,
   In the toilsome track of care:
While the hands are listless lying
   Shunning e'en the slightest tasks,
And the voice in feeble accents
   Help to bear a burden asks.

Break no reed God's hand is bending
   Quench no fire nearly gone,
But with hope the soul sustaining
   Lead the fainting spirit on.
Point to heights above the trials,
   Bathe the worn and aching feet,
Whisper words of love and kindness
   Comfort thou with influence sweet.

TRIOLET.

IF I could be a sunbeam, dear
   How bright I'd circle you around,
There should not be a pathway drear
If I could be a sunbeam, dear
Came there an ill to compass e'er,
   Before it should my smile be found
If I could be a sunbeam, dear,
   How bright I'd circle you around.

ANCHORED.

OVER the sea she sailed so grand,
My ship that sailed for Fairyland.

Her keel was silver, her prow was gold,
She freighted full on deck, in hold,

With gems that shone in morning light,
And gleamed like stars at evening bright,

But ah, she came not back again;
I waited long, but all in vain.

I wept at first as weeks went by,
The tears that never seem to dry.

I could not bear the goading thought,
My jewels all are gone for naught.

*   *   *   *   *   *

At last the silent wings drew near,
And angel voices I could hear:

So low, so sweet, I knew not earth
Had power to give such music birth.

They held my spirit till its grief,
From passion's spell had found relief.

And this is what they said to me:
"Your treasure is not lost at sea.

Though no more comes she back to you,
One day you'll cross the waters blue,

To find your precious gems re-set,
And gleaming brighter, fairer yet.

*   *   *   *   *

And then alone I was once more,
To wait upon the hither shore.

With restless feet, the sand I pressed,
As if to see which way was best.

But all was dark: I could not see
The paths that God had marked for me.

He wills that we should trust in him,
In sunshine bright, or shadows dim.

And now again I gather up
Bright sparkling gems to fill my cup.

And when another ship I see,
Be there a place or not for me,

With treasure fair I'll freight again,
And send it o'er the mystic main.

TWO LITTLE HANDS.

TWO little hands, all dimpled, pure and fair,
  As if the angels crushed the rosebuds there;
Two little hands that seem too small to clasp
The lightest burden life holds in its grasp;
And yet these little hands must larger grow
And all through life must woman's portion know.
I wonder which, could it before them lie
To choose the one, and pass the other by,
These little hands will take; will joy appear
When they shall banish self for duties near?
For parents, kindred, those they truly love,
Rise all the trials of the world above?
Bend down o'er suffering's lonely, dreary cot
And comfort give to those who have it not;
The cup of water with kind words of cheer
Bestow upon the thirsty stranger here?

Or will they find in weaving flowers bright
To wear beneath the sunbeams golden light,
To deck in raiment rich, and jewels fine,
And hold life's chalice of enchanted wine?
To find in self their highest mission here,
Nor strive to labor in a higher sphere.
Ah no, it can not be! our baby's hands
Must glean in higher fields, in broader lands,

Though thorns may tear, and brambles bar the way.
Must glean the grain, and throw the chaff away.
Each higher thought, each action good or bright
Shall shed around our darling greater light.
And yet I would not ask that flowers few
Her path of care to womanhood should strew ;
I fain would see their beauty frequent spring,
And all their fragrance o'er her pathway fling.
And see her free, their sweetest blooms to twine
With those Christ's spirit makes, on earth, divine.

HAVE FAITH.

WHAT though the way looks long and cold,
    No rays of light o'erhead enfold:
No loving hand thy steps to guide,
No flower thy path to grow beside,
      Have faith!

O, listen thou, and hear that word,
Which through the darkest hour is heard.
And know that He who rules above
Protects us all with perfect love.
      Have faith!

And if 'tis best the clouds will clear,
The sunshine through the rift appear;
Sweet flowers bloom afresh for thee
O, doubt it not, 'twill surely be.
      Have faith.

I walked a weary way and long,
Where scarce awoke a bright bird's song;
And flowers fair that opened sweet
All withered were beneath my feet,
      Yet faith

I kept; and though sometimes came doubt,
With restless feet, to stamp it out,
I trimmed again and gave more oil,
My meed of yet more earnest toil,
    To faith.

And now, through faith itself, I see
The silver rift that breaks for me.
I see the heights, their caps of blue;
I hear the bells in concord, too,
    While faith

Shines brighter, like the northern star,
And says; "The resting is not far,"
While nearer sounds that ringing bell,
Repeating this: "All's well! all's well!"
    Have faith.

BE TRUE TO ME.

BE true to me! Yes ever 'tis love's plea
    From out of all life's richly garnered store
  Take I this grain, I would not cast it o'er
The land, nor freight with it a bark for sea,
Though others pave the way with flowers for thee,
    Blend smiles with speech from golden chalice pour
    Like wine rare thought till thou canst quaff no more,
I still will cry, dear heart be true to me!
O poor were I, a beggar through the land
    Bending at every door for daily bread
      Gave I that human wish so fond, away;
    'Tis selfish? yes I know, but through life's day
  Angels walk not; I have no wings to spread;
My heaven is here—the lot wherein I stand.

MY BEAUTIFUL STAR.

I SAT last eve and watched the distant sky,
  The sombre clouds came one by one, and stretched
Themselves along until they seemed a sad
Funereal pall. Then suddenly appeared
A little rift within a cloud, and slow,
And still, as if it feared to meet the gloom
A tiny star came stealing out. So small,
It seemed at first a speck of light, that e'en
The feeblest breath of passing breeze might kill.
But gradually it larger grew, more bright,
And cast its cheery beams on all around.
The winds came up, and swept unkind and rough
Across my little star but still its light
Ne'er changed. The clouds grew darker, heavier yet,
I wondered will they fright it back beyond
My gaze, or will it wait the gathering storm
To pass, then shine for me with fairer light?
I watched, and soon a gleam, a tint so faint,
But yet so true, like love's first kiss, upon
The dear one's lips, just touched the cloud's dark edge,
I knew my star was safe. The dark-robed host
At last were gone! And there before my gaze
As fixed as when my eyes first greeted it
With light more strong, with smile more sweet for clouds

My Beautiful Star.

O'ercast, my star appeared to gladden me.
Around were other stars, while gathering still
They came, a train, to gild the court of Heaven!
But she, my own, my pride stood queen of all!
The one to me most beautiful! E'en so
Above my life, I watch a little star!
At first but faintest ray, but O so pure
I could but love and long to make it mine.
Again the winds blew cold and rough, the clouds
Had circled all around, but rifting through
The gloom, and clothed with sweetest, rarest, smiles
A little star shone forth to gladden me.
There may arise, and o'er me shine such lights
As grace the darkest pathways here: among
Them all, but *one* I'll see: but *one* for me
On earth will evermore be beautiful!

THE TRAMP TRANSFORMED.

I STRAYED one time where the forests leaves
   Were fallen along the way,
Where close by the side of a murmuring stream,
   A village awoke one day.
While tarrying there I beheld a tramp
   To the barn he came and went,
With shabby clothes, and a straw hat torn,
   He seemed to be content.

His neck-band harbored no collar white,
   His hands were of kid gloves bare
While the sun came down with patent brush
   To paint the brown tints there.
The young scamp laughed with a merry laugh
   "What do I care for it,
Think you store clothes, and stove-pipe hat
   For work of farm-life fit?

"And what would I do with a collar white
   Down at the pond to-day,
Oh, where would have been my broadcloth coat,
   Like bread on the waters, hey?
I've another hat," he snatched from a peg
   A bigger and broader brim,
"I wear sometimes wear when der sun is hot
   What does yer tink mit him?"

The Tramp Transformed.

But ah! to the table came one day,
    One dressed in courtly grace
In fine spun coat, and golden chain,
    Sat in that young tramp's place,
A collar now to his bosom's front
    Lent beauty and cleanly hue,
The wristbands fastened with fine-cut gems
    Gave grace to brown hands too.

The face that looked from the pleasant frame,
    The voice in its gentle tone,
The laugh that rang as we made our bow—
    'Twas the young tramp's very own.
Though honors now with the clothes he wore
    To us it mattered small:
The genial heart of this jolly tramp
    Dwelt still in Professor———

A LESSON FROM A CLOTHES-LINE.

I SAW some clothing yester-noon
   Hanging the line upon,
And as I looked the thought occured:
   "That wash is poorly done."
It looked so gray beneath the clouds
   That gathered thick and low,
They seemed to o'er each garment there
   Their gloomy shadows throw.

E'en as I gazed a puff of wind
   The wheel sent whirling round,
And soon within a sunny sheen
   A tranquil resting found.
As if spread o'er by falling snow
   Before my carping sight,
The clothes so late of dusky hue
   Appeared of purest white.

I stood and pondered thus: If we
   Contrast the clouds and sun,
We find its just the same with life,
   There's much injustice done.
We will not wait the wheel to turn
   And bring out color true,
But pass our judgment hastily
   On that by chance we view.

SHADOW WATCHING.

I SAW one sitting, thinking, thinking,
   'Neath the moonbeam's misty light
And I saw the shadows passing,
   Passing, all before her sight.
One there came from childhood's bower
   Dripping o'er with many tears
And her heart was quicker beating,
   As she thought of bygone years.

As she mused there sad and lonely
   Asking in her heart release
From the sorrows that enthralled her,
   Came unto her spirit peace.
And I saw the sunbeams wreathing
   'Round the shadow of the heart
While the cloud of sorrow bursting,
   Scattered tear-drops far apart.

And she trod the path of pleasure
   Where the roses sweetly bloom,
While around her sunshine floated
   Bearing on its sweet perfume.
But the sky again was darkened,
   And the moon a veil hung o'er:
As another deeper shadow
   Than the by-gone passed before.

I could hear her quick heart-beatings
   Through the heavy midnight gloom:
I could see her spirit-struggles
   As she paced her lonely room.
And I saw her bow in anguish,
   Listened to the tear-drops fall,
Then I saw God's waiting angel,
   With his white-wing dry them all.

And he whispered words so gentle,
   Such as angels only can,
How a shadow often brings us
   Our dear father's wisest plan.
How the rod we deem a torture,
   Wielded is with kindest love,
Often raised in benediction
   By the father's hand above.

Then the angel floated upward
   Through the azure space away,
Free to gild another shadow
   With God's smiling sunny ray.
And again the earth looked brighter
   And the flowers she could see,
So she rose and walked on trusting,
   On toward the land To Be.

Entered once that mystic region
   All along it duties grew,
There she staid to glean them gladly,
   Wove them into pleasures true.

Often o'er her lowly bending
    Hung the angry frowning cloud,
But she knew that God upheld it
    And His smile it did enshroud.

So she gleaned in light and shadow,
    Whereso'er her footsteps fell,
Ever heard the voice within her
    "God would have his work done well."
And she knew when years were ended
    And the paths of life were trod,
That her task would be accepted,
    On the balance sheet of God.

MINDFUL OF LITTLE THINGS.

IT is the little things of life
   That we should bear in mind
If we would have a garb of love,
   About us closely wind.

Sometimes, a blossom from a plant,
   A spray of flowers rare,
Has raised a storm-bowed spirit up
   And soothed away its care.

A trifling gift I've known to please
   To cheer a dying child,
For while the trembling fingers clasped
   The pale lips sweetly smiled.

And but a hand passed o'er a brow
   When pain was holding sway,
I know has stilled the throbbings all,
   And smoothed the prints away.

A silent kiss was found to hold
   The wealth no king could buy;
A pressing cheek contained a mine
   Of precious sympathy.

And tears! all know, that for our grief
   Wherever they are shed,
They bind with bands unbreakable
   The living and the dead.

The larger things are bought with price
   From earthly coinage made;
For little things the Heavenly mint
   Keeps count of what is paid.

TWO SHIPS.

I SAW two ships on a wide, wide sea,
  Go sailing along in company.

The one looked strongest to brave the tide
Smaller the one by the other's side.

Proudly they sailed, while mast and spar
Shone in the sun like gold afar.

The smaller looked safe by the larger's lee
To breast the waves of the wildest sea.

But soon the larger ship veered away,
Leaving the small one alone to stay.

And other ships too as staunch and strong,
Close in its wake went riding along.

But the smaller there with her wings of white,
Still slowly sailing, kept ever in sight.

The shadows dance o'er her shining prow
The wild winds sweep o'er her bulwarks now.

The waters rise to her high head-light,
And the land lies often out of sight.

Two Ships.

On, on, she sails o'er the waters wide
O'ercoming the shadows winds and tide.

And into the haven rides at last
When all the dangerous shoals are past.

I pray God help that larger ship strong
And guide it through peril safely along.

But I pray God's love be overthrown
That little bark that sailed all alone.

AT REST.

"There is no flock however watched and tended,
  But one dead lamb is there.
There is no fireside howsoe'er defended,
  But has one vacant chair."

SHE has passed from our earthly vision
    Has crossed o'er the mystic stream
And we sit 'mid the shadows of sorrow
    And say "it seems like a dream!"

Like a dream since she came amongst us
    With her bright sunny smiles—a bride
And asked of our love to bless her
    And a place our hearth-stone beside.

O, then gladly we gave her a welcome,
    So fair was her spirit and bright
Not a cloud but fled from her presence,
    Not a burden but quickly grew light:

And how fondly our heart-strings entwined her
    The future-time coming so near,
While impatient we waited for footsteps
    Not thinking two angels to hear.

At Rest.

Close folded in snowy white raiment
  We gather the lamb to our love
While safe with the Heavenly Shepherd,
  The mother looks down from above.

THE POET.

IT is his! the glorious summer,
   Its flowers blooming fair
And his the wild bird's warble
   That wakes the silent air.
His too the cloudlet sailing,
   Through all the azure deep:
And his the bright stars watching
   Where loved and absent sleep.

And his the calm blue ocean,
   The rough waves angry foam
The rugged mountain's fastness
   The soaring eagle's home:
For him the rosy morning
   The twilight's later hour
To charm his poet fancy
   Use all their magic power.

All these to him have language
   In low and soothing tone
To him are ever speaking
   The words to earth unknown.
All these are God proclaiming,
   His greatness and his love
All these the poet leading,
   To clearer heights above.

The Poet.

His life is wreathed with beauty,
    As fairest garden sweet,
His path bears oft the impress,
    Of sinless angel's feet
When wakes his soul in music
    His choicest songs are sung,
'Tis when the chords transcendent
    By angel hands were strung.

TISSUE TRANSCRIPT.

JUST a bubble ! I took some soap,
    Passed it the water through,
Then with a common pipe—for sport,
    Some bubbles simply blew,

The sunshine quickly filtered down
    To write in language rare,
These lines upon the tissue thin,
    And left its transcript there.

'Twas this : Soap-suds the story tells
    Of vexing washing-day,
When all who in its shadow live
    Are set the crooked way.

But look at me with lenient eyes,
    Impart your genial life,
And see with what a wealth of joy
    My homely form is rife.

There walk within the world's wide ways
    What plain as soap-suds seems,
Because no loving hand has sought
    To find the beauty-beams.

No tender touch has tried the spring
   That opes the silent lid,
And shown unto the world's wide gaze
   The hoard of treasure hid.

Then judge us not from circumstance ;
   Our diamond-colored spheres,
Are emblems of the beauty which
   In all God's work appears.

SONNET.

GO forth, O Love, and give of thy great store!
 Stand not as beggars stand, before the gate
Where princes dwell ; but throw thy mantle o'er
 Thy royal robes ; and e'er the dawn, or late
Beneath the dews, distribute of thy gold.
 Ask not for recompense ; wait not, but stray
As one who feels assured, yet not too bold,
 That he goes through an olden well-known way ;
Know thou, as all should know, who gave so great
 A gift, know too, one day he will require.
Nor giv'st from out thy gentle, pure white hand
 As earthly masters ask, the worth for hire.
Then go ; and shouldst thou meet upon thy quest
A heart bowed down, oh pillow on thy breast !

GRANT.

SOLEMN and sad on the outward wind
   My thoughts went from me to-day
To that mighty crowd where wearied, wan,
   The pride of our nation lay.
I saw the brave soldier dowered with fame
   And honor at home and abroad,
Who saved us our flag, and ensured us our "peace,"
   Escaping both bullet and sword.

And now from the foe whom none can evade,
   Who ventures past honors and fame,
To homes of the people, the little lone cot,
   For chieftain and children the same;
There comes no escape, though loved ones may weep,
   The nation in arms may arise.
The word has been said ; our hero must go,
   No ransom will come from the skies.

But memory waits on her evergreen bank,
   Entwines both the laurel and rose,
The one for our love, the other for fame,
   Wreaths over his monument throws.
While down through the years as history writes,
   Her annals in grandeur are bound,
There'll not be a name more honored of men,
   Than "Grant," where her record is found.

WITCH HAZEL.

IT was a pretty thought, sweetly exprest,
   For you that other morn, to me, to say—
  While walking through Life's autumn fading day
"Like the Witch Hazel; you are surely blest,"
Although by winds of autumn sorely prest,
   'Witch Hazel' holds her own, to cheer the way:
   Resists each effort on the earth to lay,
While bearing fruit and blossom on her breast.

How sweet to think, where nearing winter's snow,
   The heart can still keep warm; that summer's bloom
May slowly fade. Though autumn winds are cold
   Within, we can if willing so, keep room
For Love to come, fond arms around us fold;
So when the white flakes fall we hardly know.

AN INVOCATION.

BLOW softly o'er his grave, ye winds,
 And softly song-birds sing,
The sweetest fragrance in your cups,
 Ye flower-fairies bring.

And only gentlest music with
 Your jets of silver spray,
O charming, cooling fountains, through
 The months of summer play.

And thou, O blessed sunshine, bring
 The alchemy of love!
And break the clouds that gather with
 Thy beauty from above.

And, O ye stars, so faithful, with
 Your holy, truthful eyes,
Keep watch when all are sleeping, from
 Your stronghold in the skies.

BESSIE.

ONE more little floweret blooming,
   One more blossom born of May,
Two more little feet to wander
   Through this life's mysterious way.
Two more little ands to struggle,
   Knowing not the good from ill :
Two more eyes to see the shadows,
   For the tear-drops oft to fill.

Teach us, Father, how to guide her.
   Keep the little hands from stain ;
Veil the eyes from darkest shadows ;
   Bind the little feet from pain.
Lead her all the journey gently
   In the sunshine of thy love,
Fold her in thine arms when weary,
   Give her rest at last above.

RONDEAU.

O, LIGHTLY touch my heart's frail strings,
    'Tis then the music wakes and rings,
Until my being seems to thrill
With thought too deep for words to fill.
Wouldst hear such songs as running rill
Gives out when winds through woods are still,
    Poet that soothes us while he sings,
    And peace to throbbing bosom brings.
        O, lightly touch.

Harshly? My Muse will spread its wings
And soar afar from earthly things.
    Wouldst wake to answer at thy will
    Those silent cords to banish ill ;
Then while to life the music springs,
        O, lightly touch.

FATHER, GUARD MY BOY.

KEEP him, Father, from all evil,
    Guard him lest he go astray,
Fold him in thy loving-kindness
    All along his onward way.

Help him bear the heavy burdens
    Lest they bend his spirit low,
Gild the cloud above his pathway
    With Thy precious promise-bow.

Lead him where his work awaits him,
    Strengthen thou his ready hand,
Till amid his life's hard battle
    He may like a hero stand.

God of the widow! Father listen,
    While she bends the suppliant knee,
And her heavy weight of sorrow,
    Brings by faith and trust to thee.

Earthly friends too often fail us,
    But the "Rock of Ages," thou;
And on thee in doubt and darkness,
    Lean I in my weakness now.

MY FLOWER BOAT.

I LAUNCH my Boat at Easter tide
   To try the perilous main;
I can not tell, I do not know,
   If e'er the port to gain.
If she shall skim the waves, a bird
   With pinions fleet and free;
Or if her white sails drag the deck,
   A wreck come back to me.

She bears bright hope above her prow,
   Her helm strong faith doth hold;
Her sailors all are tried and true,
   To brave the wild winds, bold.
So well equipped I think will pass
   The storm-king safely by,
Until she anchors where o'erspreads
   A blue and cloudless sky.

Her freight is light; no heavy ore
   Between decks stows away;
I load her but with simple flowers,
   I gathered day by day.
Perhaps some heart that lowly bows
   At sight of her will cheer;
Before those silent messengers
   Some pain may disappear.

And so I trust my little boat.
　　Let times be good, or ill,
I think she'll reach the destined port
　　Her mission to fulfill.
But if she grate upon the rocks,
　　Deck, spars, and sails be torn,
Still floating on, her precious freight
　　Will to some shore be borne.

There fairer flowers than aught of mine,
　　From tiny seed may spring;
While opening neath a brighter sky
　　Shall surer solace bring.
Then, if I may not have the joy,
　　I thought were mine to gain,
My venture at this twilight hour
　　Tempts not the tide in vain.

MIRIAM.

"Standing where the brook and river meet."

THERE the angel saw the maiden—
    Saw the river's outward flow,—
Ships that were with trial laden
    O'er the waves go to and fro.
Knew he all the sorrow waiting
    Those who brave the billowy tide;
Where the rocks lie deeply hidden,
    Where are wrecks the shore beside.

Softly folding down her eyelids
    Ere life's sunshine died away—
Bore he that sweet, gentle maiden
    Into realms of endless day.
Loving hearts so sorely stricken
    See ye not that radiant star?
Christ was born on Christmas morning!
    Heaven is not an island far.

All along Life's onward journey
    O'er each path that star shall gleam
Brightest, where the darkest shadows
    Closing in around shall seem.

Falling on the shores of Memory
Never there to glow in vain!
Though the years may mark their changes,
One *sweet face* will *true* remain.

DETROIT, DECEMBER 23, 1894.

www.ingramcontent.com/pod-product-compliance
Lightning Source LLC
Chambersburg PA
CBHW031349160426
43196CB00007B/783